THE CURRENT WAR

A Battle Story Between Two Electrical
Titans, Thomas Edison And George
Westinghouse.

2nd Edition

Adam Cline

Table of Contents

Introduction..5

Chapter One: The Wizard of Menlo 6

Chapter Two: George Westinghouse 18

Chapter Three: Nikola Tesla, Father Of The Electrical Age 28

Chapter Four: AC vs. DC – The Pros and Cons 37

Chapter Five: The Currents War Gets Muddy.................45

Chapter Six: The Aftermath of The Current Wars...........61

Conclusion 68

Columbian Exposition--the grand court at night--
electrical illumination of MacMonnies's fountain and the
administration building. Halftone illustration by Charles
Graham, 1893

Introduction

In the late 1880s and early 1890s, the introduction of electricity brought with it two competing systems of electric power transmission. A powerful individual backed each system. On one side was Thomas Edison, the savvy inventor and businessman. On the other side was inventor and industrialist, George Westinghouse. The two of them got embroiled in a nasty confrontation as each of them fought to ensure his system would become the industry standard. In this book, Author Adam Cline gives a fascinating account of a commercial and technological feud that involved a public debate over the safety electricity, an aggressive and deceitful propaganda campaign and the introduction of the electric chair. Read on to find out what it would take to win the war of currents.

Chapter One: The Wizard of Menlo

The industrial and technological history of America is characterized by bold, daring and insanely creative inventions. However, for the inventors, the greatest challenge in bringing their inventions to life wasn't beating the technical odds. The biggest challenge lay in triumphing over the competition. Each of America's iconic innovators had an equally ambitious and determined rival. This set the stage for dramatic rivalries that were characterized by deceit, fierce power struggles, and raw ambition. The limits of human genius were tested as each innovator fought to overcome their equally brilliant, cunning and ruthless rivals to turn their ideas into reality.

In the early 1900s, it was the Wright brothers who were involved in a bitter rivalry with Glenn Curtiss in the race for flight, to the extent of putting their lives at risk. Before this, inventors Samuel Colt and Daniel Wesson engaged in legal battles as they sought to create the perfect revolver and, therefore, change the course of how wars are fought. In the 1940s, physicists Robert Oppenheimer and Werner Heisenberg went head to head as they strove to harness the laws of physics to create the atomic bomb – the world's most destructive weapon at the time. Perhaps, the most famous rivalry in America's history is that between two of America's most brilliant minds, Steve

Jobs and Bill Gates. The two engaged in a heated battle as they sought to dominate the new information age and bring the personal computer to every home and office.

Long before the above rivalries was another bitter rivalry that has had a significant influence on modern life as we know it. More than a century ago, Thomas Edison and George Westinghouse were embroiled in a nasty battle in their quest to make electricity an everyday convenience for the masses. Thomas Edison championed the direct current system while Westinghouse advocated for the use of the alternating current system. Knowing that there was room for only one of these two systems, Edison set in motion a great legal, political and marketing game to undermine Westinghouse's alternating current system. In so doing, Edison paved the way for a long and vicious battle that was fought on the front pages of newspapers and in the American Supreme Court, even leading to America's first attempt to use electricity for human execution. Where did all this start?

It all started with a gentleman named Thomas Edison. Thomas Edison was a savvy inventor and businessman who is considered to be one of America's greatest innovators. He is one of the people who helped shape America's economy in the late 80s and early 90s. Over the course of his life, Edison would file over 1000 patents.

As a young child, Edison attended public school for only 12 weeks, where he was a hyperactive kid who was prone to distractions. His teachers deemed him to be difficult. His mother, who was a teacher herself, pulled young Edison from school and started schooling him at home. From a young age, Edison had a voracious appetite for knowledge. He read a lot of books on various subjects. This cultivated in him a culture of self-education and learning that would have an enormous impact on the rest of his life.

Edison got into entrepreneurship and inventing early in life. At the age of 12, he left school and started selling newspapers, fruit, and snacks to passengers on trains along the Grand Turk Railway Line. Since he had access to the news bulletin sent to the station office each day, Edison started printing his own small newspaper, which was an instant hit with the train passengers. This was the first in a long string of entrepreneurial ventures that Edison would undertake in the course of his life.

While selling the newspapers on the trains, Edison had access to the train baggage car, where he set up a small laboratory to conduct chemical experiments. During one of his experiments, some chemicals he was working on exploded, and the car caught fire. After the incident, he was kicked off the train, which forced him to start selling his wares at the different stations lining the route.

During his time working on the railroad, a near-fatal event turned into a blessing for young Edison. A three-year-old kid was almost run over by an errant train, but Edison saved him. As a show of gratitude, the child's father taught Edison how to operate a telegraph. By the age of 15, he got employment as a telegraph operator. Over the next five years, Edison roamed all over the country working as an itinerant telegraph operator, taking the place of operators who had joined the civil war. All this time, Edison was still an avid reader, studying and experimenting with various aspects of telegraph technology. He also familiarized himself with electrical science.

In 1866, when he was 19, Edison went to work for the *Associated Press* in Louisville, Kentucky. Here, he worked on the night shift, which gave him enough time to read and experiment. Since he lacked in formal schooling, he developed his own style of thinking and inquiry, conducting thorough examination and experimentation until he could prove facts. Early on, Edison had developed a hearing problem. However, this did not hinder his performance as a telegraph operator since, at the time, Morse code was usually inscribed on paper. As the technology advanced, receivers started using a sounding key, which made it difficult for Edison to operate the telegraph. This left him without a job.

At the age of 21, he returned home, only to find his father without a job and his mother developing a mental illness. Edison had to do something to take control of his future. On a friend's suggestion, he ventured to Boston, which was the center for science and culture at the time. Here, Edison felt at home. He landed a job working for the Western Union Company. During this time, Edison invented and patented an electronic voting device, which would allow votes in the legislature to be tallied quickly. However, the lawmakers in Massachusetts were not interested in the device, and it became a flop.

In 1869, at the age of 22, he left Boston and relocated to New York City. Here, he designed and developed the Universal Stock Printer, a device that helped to synchronize multiple stock tickers' transactions. He sold his invention to The Gold and Stock Telegraph Company for $40,000. With his new-found wealth, he quit his job as a telegrapher and fully focused on being an inventor. He set up a research facility at Menlo Park, New Jersey.

Soon after setting up the research facility at Menlo Park, Edison came up with a device known as a carbon transmitter. This device made it possible to transmit acoustic signals with more clarity and at higher volumes, thereby improving the audibility of the telephone, which had been invented by Graham Bell. That same year, while trying to make improvements to the telephone and the telegraph, Edison invented his most original device at the time – the phonograph.

This device made it possible to record sounds by making markings on a sheet of paraffin-coated paper. The sounds could be played back by moving the sheet of paper beneath a stylus. At the time, the telephone was seen as the acoustic version of the telegraph. Edison was trying to come up with a device that could transcribe messages in human voice, just as they were received. After introducing the phonograph, it became an instant hit, and though it would take another decade before it would be commercialized, it earned Edison the title, 'The Wizard of Menlo.'

In 1877, during an expedition with several other scientists, Edison and the other scientists entered a discussion about how practical it would be to 'subdivide' arc lights, which were very intense. By subdividing them, the arc lights could be used in place of the small, individual gas burners that were popular at the time. The problem with all existing bulbs was that they burnt out as a result of overheating. Edison thought that this was a problem he could solve, and in front of the group of scientists, he made the bold announcement that he would come up with an inexpensive, mild, and safe electric bulb that would replace gas burners.

Thomas Edison with the engineers and technicians of his Menlo Mark workshop. Edison is under the central arch, leaning against the support with his hands in his pockets.
Ca. 1880.

In 1878, Edison set out to create a safe and inexpensive form of electric lighting to replace the gaslight, which was the most conventional lighting method at the time. Various scientists had been grappling with this problem for the last 50 years. However, all existing versions of the light bulb had some faults that made them economically impractical. Some had an extremely short lifespan, and others required an extremely high electric current while others were costly to produce. Edison was determined to come up with a cheap and efficient light bulb that could be commercially applied on a large scale.

Early Light Bulbs

Years before Edison arrived on the scene, other scientists had been experimenting with light bulbs. The first documented attempt at making the light bulb was in 1800. An Italian inventor by the name Alessandro Volta came up with the voltaic pile, the first ever practical method of electricity generation. His invention was composed of alternating discs of copper and zinc. In between these discs were layers of cardboard soaked in salt water. When a copper wire was attached to both ends, the setup conducted electricity, while the copper wire emitted some glow. The voltaic pile was the precursor to modern day batteries, while the glowing wire is viewed as the first ever manifestation of incandescent lighting.

Shortly after Volta showcased his invention to the Royal Society in London, an English chemist and inventor by the name of Humphry Davy connected charcoal electrodes to the voltaic pile, thereby creating the first electric lamp in 1802. Davy's invention was known as the electric arc lamp since the carbon electrodes produced a bright arc of light between them. While this was a significant improvement on Volta's glowing copper wire, Davy's electric arc lamp was not very practical. This is because it couldn't produce light for long. In addition, it was so bright that it could not be comfortably used in an indoor setting.

For the next couple of decades, several other inventors came up with their own versions of light bulbs, though none of these was commercially applicable. In 1840, Warren De la Rue, a British scientist, came up with a design that consisted of a vacuum tube enclosing a coiled platinum filament. Passing an electric current through the vacuum tube generated a glow of light. Warren De la Rue surmised that the high melting point of platinum coupled with the lack of air in the vacuum tube would make the bulb durable. This bulb design was actually quite effective, but its commercial viability was impractical because of the high cost of the platinum filament.

In 1848, another Englishman named William Staite came up with an improved design of Davy's electric arc lamp. To allow it to burn for longer periods, William developed some clockwork mechanism whose function was to regulate the movement of the electrodes, thus improving their durability. Unfortunately, William's lamp depended on an expensive set of batteries, which prevented this lamp from being produced commercially.

Two years later, in 1850, Joseph Wilson Swan, another Englishman, came up with a light bulb that utilized carbonized paper filaments enclosed in a vacuum glass tube. By 1860, he had come up with a working prototype. However, Swan lacked a good vacuum pump and an adequate supply of electricity. This reduced the lifetime of his light bulb, making it ineffective. In the 1870s, with the availability of better vacuum pumps, Swan continued trying to improve his bulb. In 1878, he came up with another bulb that lasted much longer than the first one. This one relied on a treated cotton thread. The cotton thread had longer durability and prevented early bulb blackening, a problem that was common in his first bulb.

While Swan was conducting his experiments, two Canadian inventors, Henry Woodward and Mathew Evans, were also trying to develop their own light bulb. In 1874, the two filed a patent for their first electric lamp. This one used two electrodes that held carbon rods of different sizes in a glass cylinder containing nitrogen. They tried to produce their invention commercially, but unfortunately, they did not meet any success.

By this time, Edison had started looking into how to create a safe and affordable electric bulb. He looked at Swan's design and came to the conclusion that the only problem was the filament. Edison realized that the solution was to use a thin filament with high electrical resistance. Such a filament would only need the tiniest amount of current to make it glow, making it practical. In 1878, with financial backing from the Vanderbilt family and J. P. Morgan, Edison began serious research and development on the light bulb. He also bought the patents of Henry Woodward and Mathew Evans. In December 1879, he demonstrated his improved light bulb.

Edison continued to conduct tests on various types of metals for the filaments. Several months and experiments later, Edison discovered that he could make his light bulb last for over 1200 hours if he used a carbonized bamboo filament. This discovery made it viable to commercially produce light bulbs. In 1880, Edison founded the Edison Electric Light Company and began mass production of his bulb and began marketing its widespread use.

For Edison, inventing the light bulb was only half the battle. To compete with the gas utilities, which were used to provide lighting at the time, Edison would need more than the light bulb. He would need to come up with a whole new industry. Remember, at this time, electricity was still a strange concept. There was no wiring to get electricity from the generation plant to people's homes. There were no plugs and sockets in people's homes. If people were to adopt the light bulb, Edison would need to come up with an entire system to get power to the people.

Not one to be easily stopped by challenges, Edison set out to create an electric utility that would enable him to distribute electric power to cities all over the world. His utility was based on the direct current (DC) system. Edison filed hundreds of patents for generator components, switching stations and power regulators for his DC electrical system. In 1882, Edison's first generating station on Pearl Street was turned on, providing 59 customers in lower Manhattan with 110 volts of direct current electricity. His efforts led to a decline in the profits of gas tycoons, who started spewing propaganda to discredit his new system. To counter this, Edison made his new lighting system as similar as possible to the existing gas lamp systems. His new lights were controlled by a turning key similar to the one used in gas lamps, had an almost equal brightness and were even referred to as burners.

Even as he was building hydroelectric stations to generate DC power across the United States, Edison was quick to recognize that DC power had some limitations. Transmitting DC power over long distances led to significant loss of energy. This meant that the end users of DC power had to be located within a one-mile radius of the power generating plant.

Recognizing these challenges, Edison gave the challenge of coming up with a suitable alternative to Nikola Tesla, a young Serbian engineer and mathematician he had recently hired. Just like Edison, Tesla was a genius. Tesla was quick to point out that the future of electric distribution was in alternating current. However, Edison immediately dismissed Tesla's ideas, terming them as "splendid but utterly impractical." Feeling crushed that Edison did not recognize his ideas and after a dispute about compensation for his work, Tesla left Edison and set out to build his own alternating current motor and raise funds for Tesla Electric Light and Manufacturing, a company he founded soon after leaving Edison. This move would later come to haunt Edison.

Chapter Two: George Westinghouse

Edison was not the only person thinking about lighting American homes with electric power. At about this time, George Westinghouse, another American inventor and industrialist was also making headway into electrical production and distribution. Over the course of his life, Westinghouse formed more than 60 companies and filed over 300 successful patents of his inventions.

George Westinghouse was born in 1846 in Central Bridge, New York. During his early childhood, his school teachers considered him a failure. He was a big-bodied, clumsy boy with an uncontrollable temper, a strong will, and a readiness to get into fights. Whenever he got the opportunity, young Westinghouse would escape from school and make wooden engines using his jack-knife. Due to his poor performance in class, neither his teachers nor his parents were too optimistic about what would become of him. Little did they know that, like many other great men, he was an 'ugly duckling' who would turn out to be a swan.

At the age of 14, Westinghouse finally quit school. At the time, his father had opened up a shop that specialized in small steam engines and agricultural machinery. Westinghouse went to work in his father's workshop, earning wages of 50 cents a day. At the time, his father felt this was too much a wage since the boy spent most of his time trying to come up with machines to do his work instead of actually working. His father considered him a very lazy young man.

When he was 15 years old, the American Civil War began and Westinghouse enlisted in the Union Army, serving in the cavalry. After serving in the army and navy for four years, Westinghouse resigned and returned home. He enrolled at Union College, but he did not find the curriculum interesting, so he dropped out without completing his first term at the school. He went back to work in his father's shop, earning wages of $2 a day and tinkering with the machinery at the shop whenever he could. In late 1865, he came up with a rotary steam engine, for which he received his first patent. However, the engine was soon proved to be impractical.

After the rotary steam engine, Westinghouse researched and invented car shock absorbers that used compressed air. Before his invention, riding in a car was quite rough and uncomfortable. The automobiles back then relied on leaf and coil springs. Coil springs are essentially coiled pieces of metal that compress and expand as the automobile goes over bumps. Leaf springs, on the other hand, are curved pieces of metal that are placed below the body of the automobile. As the car moves over bumps, the metal pieces give a little, absorbing some of the impacts from the bumps.

Westinghouse's compressed air shock absorbers relied on pressurized gas, instead of metal, to relieve the impact of bumps. Whenever an automobile goes over bumps on the road, the pressurized gas within the shock absorbers resists the force from the bump, thus maintaining a smooth ride. To this day, many off-road and luxury vehicles use compressed air shock absorbers similar to the original ones invented by George Westinghouse.

If you take a look back through history, you will realize that many inventions were spurred by accidents. When he was 20, Westinghouse was involved in a railway wreck. While the accident wasn't fatal, the wrecked train cars blocked the line for over two hours. Ever the solution-oriented person, Westinghouse immediately thought of a device that would help railway companies get their derailed freight cars back on the tracks. He patented his device and sold shares into his patent to two investors. This moment marked the real beginning of his career as an inventor and his interest in the improvement of railway travel. His continued interest in the railroad industry would soon lead him to his first major invention.

From early childhood, Westinghouse had a knack for noticing things that held great potential. During his time in the army, he had realized that railroads would be an essential part of industrializing the nation. However, the existing trains were not safe, which held back further development. Soon after inventing the device to right derailed rail cars, Westinghouse witnessed another rail accident that spurred his first major invention.

Westinghouse witnessed a fatal head-on collision between two trains. On enquiring about the cause of the accident, Westinghouse learned that the two engine drivers saw each other, but the existing brakes were unable to stop the trains in time to prevent the accident. At this time, trains used a manual system, which required brakemen to move from car to car, manually applying brakes in each car. Convinced that this was a very ineffective way of stopping trains, Westinghouse spent some time studying the manual braking system, trying to see what improvements he could make.

In his first attempt, he tried using a long chain, but he found it to be too clumsy. His second attempt tried using steam, but he found that the effectiveness of the brakes would be compromised by atmospheric temperatures. Soon after, he read an article in a magazine about how compressed air had been used to dig a tunnel. This gave him the breakthrough he needed. He came up with a method that used compressed air to actuate the brakes in all the cars simultaneously. He formed a company to commercialize his air braking system, which was widely accepted. In 1893, the Railroad Safety Appliance Act made it compulsory for all American trains to use air brakes. Under Westinghouse's lead, the air brake system became the standard in America and Europe.

As rail traffic increased, Westinghouse saw the need for better railroad signaling system. He combined his own ideas with those from bought patents and came up with a complete electrical and compressed air railroad signaling system. He also developed the 'frog', an interlocking switch device that allowed trains to switch between two tracks.

In 1883, a well that had been drilled in his father's yard turned his attention to the natural gas industry. Here, his special knowledge of air compression brakes would help him solve major problems facing the piping and distribution of natural gas. At the time, using natural gas as a source of fuel at home was not very safe. Natural gas created a lot of pressure when leaving the well. This pressure had to be maintained if this gas was to move through the distribution pipes to people's homes. However, having highly pressurized gas leaving the gas outlets within people's homes was not safe. As a solution to this problem, Westinghouse came up with a reduction valve that allowed the gas pressure to be reduced before getting into people's homes. Within two years, Westinghouse had created several other inventions that allowed for easy control and distribution of natural gas. Using his new inventions and ideas, he formed a gas distribution company to serve the Pittsburgh area. He made natural gas safe enough to be used within people's homes, and Pittsburgh became the first city to have a widespread natural gas delivery system.

Just like he had realized the importance of rail in industrializing the country, Westinghouse also recognized the massive potential of electricity. His knowledge from working in the gas distribution industry gave him an incentive to come up with a better system for the distribution of electric power. One of his inventions was a reduction valve that made it possible for the high pressure of gas from the well to be reduced at the point of use. Westinghouse believed that a similar approach would work with the distribution of electric power.

In 1884, Westinghouse got into electricity generation and distribution when he hired a physicist, William Stanley, to help him develop his own DC lighting system. At this time, the DC system was the most common in America, and so Westinghouse followed suit. With the help of Stanley, Westinghouse experimented with and researched DC electricity. In 1886, he installed a direct current plant that powered incandescent light bulbs at New York's Windsor Hotel. Shortly after, Westinghouse built another similar plant to power up the Monongahela Hotel in Pittsburgh.

In August of 1886, Westinghouse completed the installation of his first direct current central station in Trenton, New Jersey. This plant was comprised of six, 100-volt DC dynamos. While they were installing the direct current systems, Stanley was trying to figure out alternator design principles that would make it possible to use alternating current to power up arc lighting. This new system seemed appealing and plausible to Westinghouse, and he dedicated a lot of money to developing it commercially. However, there were several hurdles along the way. The most difficult one was the fact that direct current arc lamps were very superior to the alternating current arc lamps that were in existence at the time. The alternating current arc lamps wasted much of the light that they produced, which made them impractical. They also produced a lot of noise, something that was objectionable. Because of this, Westinghouse gave up on the idea of an alternating current electric lamp. The effort was instead directed to improving direct current. However, there were still strong technical reasons that alternating current held potential as a better method of transmitting electrical energy.

Shortly after giving up on alternating current electric lighting through a publication known as *Engineering,* Westinghouse got wind of a new type of AC systems that were being trialed in Europe. These AC systems used a transformer to step up power to high voltages before distribution. This way, the power could be transmitted over long distances without the energy loss that was the bane of DC systems. Afterward, another transformer would step down the power to lower voltages that were safe for consumer use.

Being the clever businessman he was, Westinghouse immediately recognized the potential these new AC systems presented. With such a system, he would enjoy economies of scale by building large centralized power generation plants that were capable of supplying electricity to vast areas, even in areas with disperse populations. Rather than creating another DC system like Edison's, Westinghouse saw AC as his ticket to a genuinely competitive system. Since Edison's system could only reach customers within a one-mile radius of his generating plants, Westinghouse saw a huge market in the huge patches that were outside the reach of Edison's plants.

By this time, John Gibbs of England and Lucien Gaulard of France had made a successful demonstration of AC systems in London. Westinghouse reached out to Gibbs and Gaulard and imported a set of their transformers, as well as a Siemens AC generator, and began experimenting with the AC system. He hired three American electrical engineers who helped him alter and perfect the Gaulard-Gibbs transformers to come up with a constant-voltage AC generator.

By this time, Stanley's health had begun to deteriorate, forcing him to move to Great Barrington, Massachusetts. Westinghouse built a laboratory in Great Barrington, allowing Stanley to continue experimenting with alternating current and to demonstrate its practical uses. After tinkering with Gaulard's and Gibbs' transformers, Stanley, with the help of the other American electrical engineers, came up with the design of a new transformer that could be used commercially.

They built an experimental power distribution network in Great Barrington. This became the first ever multiple-voltage AC power system in America. The network depended on a hydroelectric generator that was capable of producing 500 volts of AC power. His transformers stepped this up to 3000 volts for transmission, then stepped it back down to 100 volts so that it could be used to power electric lighting. It was at this time that Westinghouse formed the Westinghouse Electric Corporation. This was followed by the installation of another commercial alternating current plant in Buffalo, New York, in November 1886.

The following year, Westinghouse installed 30 more AC current plants. However, he was still faced with a major challenge that limited the expansion of his AC system – the lack of a good AC electric motor and the lack of a good metering system. A year later, in April 1888, with the help of an engineer, Oliver B. Shallenberger, he came up with an induction meter that measured alternating current using a rotating magnetic field. The principles on which the induction meter was based are still in use today. In the same year, inventor Nikola Tesla came up with the designs for a polyphase brushless induction motor. He built and demonstrated the workings of his motor, which was also based on a rotating magnetic field.

Apart from being a great thinker and having an eye for great ideas before their time, Westinghouse also had a knack for identifying people who could help him achieve his goals and bringing them to his fold. After hearing about Tesla's motor, Westinghouse secured the license to Tesla's patent for the induction motor, as well as several of his patents on other electrical devices, including his transformer designs. He also paid Tesla to come and work with him, a move that would be his first major step toward triumphing over Edison. Apart from buying Tesla's patents, Westinghouse also bought the patents of Galileo Ferraris' induction motor. Galileo Ferraris was an Italian electrical engineer. Armed with this, and the support of Nikola Tesla, Westinghouse was ready to face the electrical giant that was Edison. Over the next couple of years,

Westinghouse and Tesla would work together toward the common goal of achieving victory over Edison and his DC current.

Westinghouse Electric began installing AC lighting systems in different parts of the country, mostly serving places with lower populations where Edison's DC power could not reach. Westinghouse was also a shrewd businessman. To undercut Edison's business and get a foothold in the cities, he started selling electricity at a loss. This quickly gained him lots of customers. Within a year, his generating plants were more than half the number of those owned by Edison. With his determination and enormous capital, Westinghouse was slowly gaining market share, and Edison took notice. Soon enough, he would take action to protect his business, resulting in a nasty confrontation that came to be referred to as the "War of Currents."

Chapter Three: Nikola Tesla, Father Of The Electrical Age

Though the War of Currents was principally between Thomas Edison and George Westinghouse, it would draw in another man who would significantly affect the outcome of the feud. This man was Nikola Tesla. Nikola Tesla was an inventor, physicist, electrical and mechanical engineer and a futurist who came up with ideas that were well ahead of his time. Tesla devoted his entire life to experimenting with electricity, magnetism and radio waves. However, Tesla is best known for his contributions to the generation, transmission, and application of electric power, which would define the course of the battle between AC and DC electricity.

Nikola Tesla was born on a stormy night in 1856 in Smiljan in present-day Croatia. He was the fourth of five siblings. Tesla's father was a Serbian Orthodox priest and writer, while his mother was an inventor who specialized in small household appliances. Growing up, Tesla developed an interest in electrical inventions, spurred by his mother's inventions. Tesla would credit his creative abilities to his mother's genetic influences. In his early childhood, he joined a primary school in Smiljan, where he was schooled in arithmetic, religion, and German. After clearing primary school, Tesla went to a high school in Karlovac in 1870. In 1873, as a result of his unnatural intelligence, he graduated from the course after only three years instead of four.

After graduating from the Higher Real Gymnasium, Tesla got a Military Frontier scholarship to attend an Austrian polytechnic in Graz known as the Realschule, Karlstadt. Tesla's intention was to study advanced engineering and physics, going against the wishes of his father, who wanted him to become a priest. However, after a short while at the school, his focus and interest moved to electricity. Tesla saw the Gramme dynamo for the first time while at Graz. This dynamo worked as a generator and converted into an electric motor when reversed. This gave him an idea on how to use AC to advantage. Despite being a brilliant student during his first year, Tesla developed a gambling addiction in his second year. This addiction ruined his studies and he was unable to obtain a degree from the polytechnic. He went on to attend the University of Prague and once again left without graduating.

In his early life, Tesla had shown signs of mental illness, which would become a constant part of his life, even in old age. He would undergo periods of illness, interspersed with periods of breathtaking inspiration. In moments that he claimed were accompanied by blinding flashes of light, Tesla would spontaneously visualize theoretical and mechanical inventions. One of his biggest and most unique gifts was the ability to visualize complex designs and computations in his mind. He could design his inventions without having to write down any plan or make any drawings. Instead, he relied solely on his mental images.

After leaving school, Tesla moved to Hungary to work at the Budapest Telephone Exchange, a telegraph company. Here, Tesla worked as a chief electrical engineer. During the time he worked at the phone exchange, Tesla gained a lot of knowledge about twin turbines and helped the company come up with a device that made amplification possible when using the telephone. It was during this time that Tesla got the idea for the induction motor. One day, in a moment of inspiration, while taking a walk through a park with his friend, he visualized the solution to the rotating magnetic field, something that had been elusive to him for a while. Using a stick, he drew a diagram on the ground to show his friend how the induction motor would work. This seemingly insignificant moment would become the first in a series of events that would have a major impact on the outcome of the war of currents.

After a short stay in Hungary, he moved to Paris to work under the Parisian branch of the Continental Edison Company. Here, Tesla worked in the new electrical power utility industry, a position that would equip him with a great deal of practical electrical engineering experience. Soon enough, the management of Continental Edison noticed his advanced engineering and physics knowledge and tasked him with designing and building better versions of the company's motors and generating dynamos. A manager who had been working at the Parisian branch recommended that Tesla be brought to Edison

Machine Works, a manufacturing division of the company located in New York City.

In 1884, Tesla moved to the United States to work for Edison Machine Works, armed with nothing more than a few pennies, a letter of introduction to the famed genius, Thomas Edison and his brilliant mind. Edison was immediately impressed by Tesla's diligence and ingenuity and the two worked together for a while, improving some of Edison's inventions. Faced with the limitations of DC electricity that could effectively place a ceiling on his company's growth, Edison assigned Tesla the task of coming up with a solution to the challenge.

Before moving to the United States, Tesla had come up with the idea of developing a brushless AC motor. He saw this as the solution to the limits of using direct current to transmit and distribute electricity within cities. With the resources available at Edison Machine Works, Tesla's invention could finally come to life. Tesla presented his futuristic ideas about alternating currents to Edison, only to have them brushed aside as impractical. Edison was convinced that DC was the only practical way of distributing electricity.

It is good to note that Edison had not received a formal education and was, therefore, no mathematician or engineer. However, he more than compensated for this with his affinity for brute-force experimentation. He would try every possible option until he came up with a solution. Tesla, on the other hand, was a mathematician and engineer who could perform calculus calculations in his mind. To understand the principles of AC electricity, one needs a substantial understanding of physics and mathematics, which Edison lacked.

At one point while working at Edison Machine Works, Edison asked Tesla to come up with improved designs for some of his dynamos, promising the Serbian engineer a $50,000 bounty if he succeeded. Tesla spent months doing research and conducting experiments and finally presented a solution to Edison. When he demanded the $50,000 bounty, Edison brushed aside the promise, terming it as a joke. This greatly angered Tesla, who quit from Edison Machine Works soon after.

After leaving Edison Machine Works, he scraped by, working as a casual laborer to raise funds for his newly formed Tesla Electric Light and Manufacturing. However, his luck changed, and he found investors who were interested in his AC electrical system. He got funding for his company and got to work, testing various versions of the AC motor and filing multiple patents for various inventions based on alternating current. He finally came up with an AC induction motor that promised efficient long distance, high-voltage electricity transmission, which he patented in 1888. This innovative electric motor had a simple self-starting design that did away with the commutator. By doing away with the commutator, this new motor was able to prevent sparking and eliminate the high maintenance costs of electric motors due to the need for constant servicing and replacement of mechanical brushes.

Later that year, Tesla got invited to the American Institute of Electrical Engineers to demonstrate the workings of his new invention. One notable characteristic about Tesla is that he was very good at coming up simple but very effective demonstrations to show people what was possible with his inventions. He demonstrated the AC induction motor and impressed the engineers.

Tesla's invention and his demonstration at the American Institute of Electrical Engineers caught the attention of George Westinghouse, through engineers working for Westinghouse who were in attendance at the demonstration. At this time, Westinghouse was already marketing his alternating current system, but he still needed a viable AC motor for his power system. Westinghouse had already unsuccessfully tried filing a patent for a rotating magnetic field-based brushless induction motor. Westinghouse saw the possibilities available with Tesla's motor and was convinced that this is what he needed to dominate the market with his power distribution system.

Westinghouse negotiated a deal that saw Tesla transfer the patents for his induction motor and transformer designs to him. In return, Tesla was compensated in cash and stock as well as royalties for every AV horsepower produced by each of his motors. Also, Westinghouse also hired Tesla as a consultant and gave him his own lab. During this time, Tesla worked at Westinghouse's research labs to make further improvements to the alternating current system to power people's homes. By teaming up, Westinghouse and Tesla and their AC system were put on a colliding course with Edison and his DC system.

Electricity. Westinghouse AC generator. The world's first. Built by Nikola Tesla and George Westinghouse. Ames Hydroelectric Plant, Telluride, Colorado. 1895 powerhouse, photo ca. 1900

With Tesla working for him, Westinghouse started installing his own AC generators all over America. Since the AC system allowed long distance power distribution, Westinghouse was able to reach sparsely populated areas where Edison's system was unable to reach. Within a year, Westinghouse was steadily gaining ground against Edison, having installed 68 AC power generating plants, compared to Edison's 121 DC plants. Things were not looking good for Edison. To make matters worse, another competitor, the Thomson-Houston Electric Company, had also gotten into the business, with 22 AC and DC-based power stations already installed. To avoid lawsuits over patent conflicts, Thomson-Houston had come into agreements with Westinghouse, paying royalties to use his induction motor patents and allowing him to use their incandescent bulb patents.

Edison was rapidly losing business to these industry newcomers and especially to Westinghouse. His company was also having to deal with increased expenses due to the rising price of copper. His low voltage DC system relied on much heavier copper wires as compared to the high voltage AC systems. His sales agents were getting demoralized since they were losing out on deals with municipalities that preferred the cheaper AC systems. They also could not reach into the rural and suburban areas where Westinghouse was getting a strong foothold. Some of the engineers in Edison's company tried urging him to consider switching to AC systems, but Edison remained adamant about DC systems. According to him, the other companies were merely using AC as a means to get around his DC patents.

Edison realized that if his business was to survive, he had to do something. Seeing that AC seemed to be winning and having no technical advantages to sell his DC system, Edison resulted to using the only remaining weapon in his arsenal – propaganda. He launched a smear campaign to discourage the use of AC, claiming that AC systems were too dangerous.

Chapter Four: AC vs. DC – The Pros and Cons

Toward the end of the 19th century, electric power was largely a new phenomenon, the preserve of a few. Only the very wealthy had access to electric lighting in their homes. The rest of the population had to rely on the other inconvenient and sometimes dangerous means of lighting, such as gas lamps. However, a few inventors and businessmen had seen the potential of electricity. They were aware that it would bring about the next phase of the industrial revolution. Each of the few electricity companies in existence wanted to be the first to bring electricity to the mainstream. The benefits were as clear as day. Whoever became the first to do this would set the standards and reap enormous benefits.

This is the situation that two companies – Edison Electric Lighting Company and Westinghouse Electric Corporation – and their respective founders found themselves in. The rivalry that sprung up between Thomas Edison and George Westinghouse in the War of Currents era was as a direct result of each of them promoting a different technology for the distribution of electric power. Each believed that their system was better suited for easy and economic distribution of electricity. This meant that the first system to gain mainstream adoption would set the standards for electricity distribution. Since patents covered both competing systems, whoever lost out in this feud would lose out on the royalties associated with their patents.

The question is, why couldn't the two agree on one power distribution technology? Was one technology more superior to the other, and if so, which one? Let's start by looking at the differences between the two.

Direct current electricity is the type of electricity that we get from batteries. Essentially, direct current flows in a single direction. A DC system is designed such that one place – the source – stores a large amount of negative charge. This leaves an equally large amount of positive charge on the other end. This creates a potential difference between the two ends. When a circuit is completed, the potential differences causes the negative charge to move from one side to the other. As the charge moves, it gets depleted from the negatively charged side, which in turn decreases the potential difference. Once the amount of negative charge is equal on both sides, the current stops flowing.

Alternating current electricity, on the other hand, does not flow in one direction. It keeps reversing direction and changes polarity several times in a second. Rotating magnets create alternating current electricity near conducting wires. When the coiled wire is passed through a magnetic field, the magnetic field induces an electric current in the wire. The current then creates its own magnetic field, which repels against the magnets. To keep the current flowing, the magnets have to keep going back and forth, reversing their polarity. This is what creates the alternating current. The easiest way to produce alternating current is to use a turbine. While there are numerous ways of turning a turbine, flowing water is the simplest and the most commonly used.

To properly understand the difference between DC and AC, we first need to understand how electric current flows. When we talk about the flow of electrons, we are not talking about the actual movement of electrons from the source to the load. Electrons move at a very slow speed. For instance, the electrons in the power cable of a typical 120V AC lamp move at a speed of about 10 cm every hour (depending on the amount of current and size of the wire). This means that if the power cord is a meter long, the lamp would produce light 10 hours after you turn on the light switch, which is not practical at all. In the case of AC, which reverses direction several times every second, it would mean that the electrons would never get to the device being powered.

What actually happens when you turn on an electrical switch is that the electrons in the wire wiggle and push the next electrons, which then push the adjacent electrons. This way, the electrical power is instantaneously transmitted to the device being powered, without the first electron having to reach the device. To make this easier to understand, I will use the illustration of a tube filled with golf balls. In our illustration, the tube will represent wire while the golf balls are the electrons that conduct electricity. Imagine having a tube whose diameter is big enough to only fit a single golf ball at a time. Take a certain length of this tube and fill it with golf balls, end to end, till you can fit in no more golf ball.

Now, get a box of golf balls. This will act as the source of electrical charge. At the other end of the tube, have an empty box to collect golf balls that come out of the tube. This acts as the 'ground'. You will notice that when you take one golf ball from the box of balls (source) and push it into the tube (wire), one ball will immediately fall from the other end of the tube and into the empty box (ground). This is how electrical flow works. The ball you pushed into the tube does not have to get to the end of the tube. Instead, it pushes on the next ball, which pushes on to the next until the ball at the end of the tube gets pushed out. This way, the energy is transmitted instantaneously. As soon as you push a ball into one end of the tube, another immediately falls out of the other end. In our little illustration, there is no point at which the tube runs out of golf balls. Similarly, there is no point where the wire runs out of electrons.

Just like the wire, electrical appliances like light bulbs and televisions have electrons in them. When a switch is turned on, the electrons in the wire next to the switch push the adjacent ones, which then push the adjacent ones until the push gets to the appliance, similar to how the balls in our illustration were pushed. Once the push (energy) gets to the electrons in the bulb, the bulb lights up. It doesn't matter from which direction the push comes. Anytime they are pushed, the light bulb goes on.

In the case of direct current, the electrons flow only in one direction, from the source to the ground, through the appliance. Owing to the difference in electric potential between the source and the ground, the electrons cannot flow in the opposite direction. In the case of alternating current, there is no fixed source and ground. The polarities change several times each minute. Going back to our illustration, this would be equivalent to alternatively pushing a golf ball into each end of the tube. This means that the balls within the tube (electrons) would be pushed in one direction, and then in the other direction. Remember, I mentioned that the direction from which the push comes doesn't matter to the light bulb. If the light bulb was in the middle of the tube, it would light up constantly, even if the push comes from different directions every cycle. That it how alternating current works.

At the beginning of the electricity revolution, the standard for the distribution of electric power was Edison's direct current. The myriad of electrical appliances we have today was not yet in existence. The principal load at the time were the incandescent bulbs developed by Edison and motors. DC electricity was designed to work well with these. The DC system also enabled direct use with power storage batteries, thereby allowing the provision of backup power when the operation of the generator was experiencing interruptions. Additionally, DC generators were somewhat more economical since they could be connected in parallel, allowing smaller generators to be used during periods of light load. For easier billing, Edison had also invented a meter that measured customers' power consumption. However, this meter was only capable of working with the DC current. At the time, all these were significant advantages that had set up Edison's DC system to be the primary system of electric power distribution.

In Edison's DC distribution system, the power generating plants were connected to heavy distribution conductors, which delivered power to customers' light bulbs and motors. One of the key features of this system is that it maintained the same voltage level throughout every stage of the distribution process. A generator producing 110 volts was directly linked to a 100 volt light bulb. The minor difference compensated for the energy that would be lost during transmission. This voltage level was chosen for two main reasons. First, the filaments in the incandescent bulbs could withstand this voltage level while providing adequate lighting as to be able to compete with gas lighting, which was the primary source of light at the time. Second, it was believed that 100 volts would not pose the risk of death from electric shocks.

Despite its perceived advantages, the direct current had some limitations. First, direct current is susceptible to energy loss during transmission. The farther a direct current has to be transmitted, the higher the energy loss. This meant that Edison's DC could not be transmitted over long distances. In addition, it required very thick copper wires to minimize the loss of energy, which led to increased material costs. To counter this limitation, Edison's solution was to build power generation plants close to where the power needed to be consumed. This meant several power plants in different parts of the city and sometimes within neighborhoods. However, this solution was not very effective. First, it was very costly to build multiple power plants. It was also not very practical to build power stations in rural areas where homes were not clustered together.

Another limitation of direct current is that it cannot be easily converted to different voltages. This meant that if you had different appliances that operated on different voltages, you had to install a separate electrical line with a particular voltage for each appliance. This was impractical and expensive since it meant laying more wires.

The newly invented AC electricity, on the other hand, did not have the limitations faced by DC electricity. Early on, Tesla had realized that electrical power could only be practical if there were a way to transmit it efficiently over long distances. Hydroelectricity had already shown great potential as a source of electricity. In most cases, however, suitable water sources were hundreds of miles from the cities where electricity was needed. An effective way of getting this power to the cities was necessary. While both AC and DC experience power loss during transmission over long distances, transmitting at high voltages minimized this power loss. With AC, the current could be stepped up to very high voltages for transmission and then stepped down to low voltages on the consumer side, with minimal power loss during the transmission. This was impossible with DC. Efficient transmission meant that one large generating plant could serve several cities that were miles apart. Since the current was stepped down on the consumer side, AC also made it possible for the same line to serve several appliances of varying voltages.

AC energy was also easier to generate. The AC generators designed by Tesla used a simple design that simplified the process of generating AC electricity. This offered a great advantage compared to the more complicated process of generating DC electricity manually. Today, most DC power is generated by fuel cells, solar cells, batteries or by the conversion of AC to DC.

AC machines were also easier to operate and maintain. DC machines depended on motors that used brushes and commutators. These components required frequent maintenance and replacements. On the other hand, the AC induction motor developed by Tesla did not require brushes and commutators, which made AC machines cheaper and easier to operate and maintain.

Chapter Five: The Currents War Gets Muddy

Seeing that his system had no technical advantage to compete with the AC system, Thomas Edison turned to a powerful human motivation – fear – to sway public opinion about the AC system. When coming up with the DC system, one of Edison's goals was safety. He was meticulous to avoid any bad press that would follow as a result of customer deaths from electric shocks. Edison surmised that, with the high voltages transmitted through AC systems, it had to be more dangerous. Therefore, he set out on a PR campaign to show the masses the dangers of AC electricity. Within a few days of Westinghouse launching his first large-scale system, Edison proclaimed that Westinghouse's system would kill a customer within six months. This was the first jab in the battle of currents. To Edison, there was no safe and workable system of transmitting AC electricity.

A year later, Edison and his company sent out a pamphlet titled "A Warning from the Edison Electric Light Company" to various companies and newspapers that were using or were considering using electric power from his competitors. In the pamphlet, he stated that his competitors were infringing on his patents and therefore companies using their services risked finding themselves in a lawsuit.

He also placed great emphasis on the safety and efficiency of his system while discrediting the Ac system. To support his argument that AC energy was dangerous, Edison included in the pamphlet clippings of newspaper stories highlighting accidental deaths arising from alternating current electrocution.

Edison's arguments were not baseless. The spread of the arc lighting systems was accompanied by chilling stories of people getting killed instantly by this new phenomenon, especially unwary linemen. For instance, in 1881, a story ran in the papers about how a drunken dock worker was instantly killed after grabbing an electric dynamo. While this gave negative press attention to electricity, it also got some other people thinking about other possible uses of this mysterious energy source. One of these people was Alfred P. Southwick, a New York dentist. After seeing the fatal consequences of electricity, Southwick saw an opportunity to develop a method of using electricity to euthanize animals. He partnered with a local physician named George E. Fell to conduct his experiments, electrocuting hundreds of dogs in the process.

Soon after, Southwick published articles on how this new energy source could be used as an alternative to hanging, using something similar to a dental chair as a restraint. According to Southwick, this method was less painful and more humane compared to hanging. Some New York State politicians, who had been searching for an alternative, following a series of messy hanging attempts, took notice of Southwick's articles. In 1886, New York governor, David B. Hill appointed Southwick as part of a commission to research on this new form of execution. As part of their research, they contacted several law, health and electricity experts to find their views about their proposed method of execution. One of the people contacted by Southwick was Thomas Edison.

Initially, Edison was against the death sentence, and he did not want to be part of this new project that proposed carrying out death penalties using electricity. However, Southwick persisted, and Edison suddenly had an idea. This was a golden opportunity to hit out at Westinghouse. Edison wrote back to Southwick stating that the most effective way of executing criminals using electricity was by using alternating machines, which were principally manufactured in the United States by Mr. George Westinghouse. Soon after, in 1888, Southwick and his commission recommended that electricity be used as a means of execution and the bill was passed by the New York government.

At this time, the number of deaths related to high voltage AC arc lighting was rising, sparking off a media frenzy that strongly condemned the "deadly arc lighting current" as well as the companies that used it. A broken telegraph line transmitting AC killed a 15-year-old boy in April, 1888. A fortnight later, an AC line killed a clerk, while a lineman working for the Brush Electric Company was killed by an AC line he was working on. This negative press was doing a lot of damage to Westinghouse's company and other pro-AC companies and fanning public outrage and resentment against AC electricity.

In June 1888, an electrical engineer by the name Harold Pitney Brown joined the anti-AC camp. He wrote a letter to the New York post condemning AC electricity as the cause of the problem, terming it as inherently dangerous and damnable. According to Brown, the public was being subjected to constant risk of sudden death just so that utilities and companies could use the cheaper AC electricity. At this time, there seemed to be no connection between Brown and Edison's company. Westinghouse and his company were getting a lot of heat from these attacks, and in an attempt to defuse the situation, he extended an invitation to Edison, asking him to visit him at his lab to discuss the state of affairs in the electric industry. Edison declined the invitation, stating that he was too busy at his research lab.

On June 8, 1888, Brown went before the New York Board of Electrical Control lobbying for strict regulation of AC electricity. Among his demands, he wanted AC electricity to be limited to 300 volts, which he claimed was the safety limit of electric voltage. However, limiting AC to 300 volts would make it almost useless for electricity transmission. There were several opponents to Brown's arguments, who pointed out that there was no scientific proof that AC was the more dangerous of the two systems. Westinghouse wrote several letters to the press highlighting the number of fires that had been caused by DC equipment. He also accused Brown of being Edison's puppet, an accusation Brown denied vehemently. Other electrical engineers emerged, questioning Brown's views and expertise on electrical systems and even gave tales of how they survived from AC electricity shocks at voltages of up to 1000 volts. This new crop of engineers took an offensive stance, claiming that DC was, in fact, more dangerous than AC.

Not one to be put down, Brown set out to demonstrate that AC was more dangerous than AC. He got in touch with Edison, asking him if he could use his equipment to demonstrate the dangers of alternating current. This was a request Edison could not refuse. He provided Brown with space and equipment at his lab and even loaned him Arthur Kennelly as a laboratory assistant. With space and equipment provided, Brown embarked on his experiments. He paid local kids 25 cents to provide him with stray dogs, which he used for his experiments with both DC and AC systems. After several experiments and lots of electrocuted dogs, Brown held his first public demonstration at the Columbia College. Defying protests by the participants to stop the demonstration, Brown applied increasing amounts of direct current to a caged dog, going up to 1000 volts. The dog survived. He then switched to alternating current and electrocuted the dog at 300 volts. The dog gave out a yelp and fell dead immediately.

After the demonstration, there were claims that the dog was weakened by the direct current, which made it succumb to the lower voltage alternating current. To disprove these allegations, Brown held another demonstration four days later at the same college. This time around, he killed three dogs by treating them to 300 volts alternating current shocks. Shortly after the demonstration, Edison told a reporter that electric power from an alternating machine was capable of killing a man within a ten-thousandth of a second. Brown and Edison were confident that this demonstration would lead to limitation of alternating current installations to 300 volts. The lobbying and demonstrations, however, did not yield fruit.

Edison and Brown did not give up. They continued their research and experiments with the aim of proving beyond all doubt that alternating current was more fatal than direct current. Soon after, they invited the New York State commission tasked with researching the use of electricity for executions for a demonstration at Edison's Menlo Park research facility. During the demonstration, they electrocuted and killed several calves and a horse. Despite the fact that the animals did not die instantly, the committee was impressed by the demonstration. They approached Westinghouse with the intent to buy three of his alternating current machines. After hearing that his dynamos were to be used for state executions, he refused to sell them. Soon after, the New York state government commissioned Brown to design and build an electric chair to be used for executions. Edison made sure that Brown incorporated alternating current in designing the electric chair.

Throughout 1888, Brown continued his attacks on Westinghouse and his alternating current system. At one point, Brown wrote a column for a magazine named the *Engineer Electrical* claiming that

Westinghouse's installations had caused some 30 deaths. Westinghouse challenged the claim. After investigations by the magazine, the claim was found to be false. Only two deaths could be directly linked to Westinghouse's AC systems.

After being commissioned to come up with designs for the electric chair, Brown conducted several experiments to determine skin conductivity, electrode placement, and design as well as appropriate voltage amounts. After several tests on various animals and discussions with the committee, it was decided that the proper voltage for carrying out human executions was between 1000 – 1500 volts of AC electricity. After the commission approved this, newspapers were quick to point out that the AC power lines running above American streets had twice the amount of voltage required to kill a man. Somehow, Brown got access to a Westinghouse AC dynamo, which he used in the design of the electric chair.

Westinghouse realized that the tests conducted by Brown were biased and were meant to sabotage the alternating current system. In retaliation, he wrote to the New York Times, poking several holes into the manner in which Brown conducted the tests. He also claimed that Brown was being paid by Edison to discredit alternating current systems. Brown denied the claims. To further prove his arguments, Brown challenged Westinghouse to a rather strange and dangerous electric duel. He was willing to receive increasing amounts of DC electric shocks if Westinghouse would submit himself to similar shocks from alternating current power. Westinghouse did not take up the challenge.

The battle of currents took a new turn in May 1889. New York had sentenced a street merchant by the name William Kemmler to death by electric chair. This was to be the first human execution using electricity. Before the execution, there was lots of press attention to the sentencing and this new form of execution. Some newspapers referred to the process as being 'Westinghoused.' One of Edison's lawyers even wrote to a colleague describing the term Westinghoused as the best choice for the procedure. However, the term electrocution ended up being adopted.

William Kemmler was born to a poor family in Philadelphia, PA, on the 9th of May, 1860. He was one of 11 children. Having been born into a poor family, Kemmler did not get any gainful education and he never learned any meaningful trade. Instead, he survived by giving his father assistance in a butcher shop. In October 1888, Kemmler married a woman named Ida Prier. However, after two days of marriage, he left her and eloped with another woman, Mrs. Matilda Tripner Ziegler. The two of them settled in one of the slums of Buffalo, NY. After moving to Buffalo, Kemmler became a fruit "huckster" and gained a modest level of success in his trade. Unfortunately, Kemmler and Matilda were having constant arguments at their home, with Kemmler claiming that Matilda took the money he made and used it to have affairs with other men and women in the slums.

As Mrs. Ziegler was preparing breakfast on the morning of March 29, 1889, Kemmler came into the kitchen and confronted her about her infidelity. As the situation heated up, Kemmler walked from the kitchen and into the barn and returned with a hatchet. Without a further word, Kemmler hacked at Matilda's head with the hatchet. He kept hacking her head, shoulders, and chest until he was sure she was dead.

After killing Matilda, Kemmler left the house with his clothes covered in blood. He started moving from house to house, telling his neighbors and anyone he encountered of what he had done. Kemmler did not show any remorse for his actions. Instead, he claimed he was happy he had killed her and was even willing to hang for it. Back in the 19th century, the judicial system was not like today's where months and years pass before murder trials come to court. Just five weeks after committing the murder, on May 9, 1889, Kemmler was found guilty of first-degree murder and was sentenced to death by electric chair.

After hearing that Kemmler had been sentenced to death by electric chair, Westinghouse financed Kemmler's appeal, contributing $100,000 towards Kemmler's legal fees. The appeal claimed that such a method of execution was both cruel and unusual. The appeal was Westinghouse's attempt to repeal the electrocution law and, therefore, prevent his AC generators from being used in the execution of criminals. He knew that if the public was convinced that his alternating system was lethal, he would suffer losses amounting to millions of dollars. Unfortunately, the appeal was unsuccessful, and the court ruled that the execution would go on.

In preparation for Kemmler's execution, the electric chair was set up at Auburn Prison in a room 25 feet long and 17 feet wide. The room would also hold 27 people to witness the execution. In another part of the prison, 1000 feet away from the execution room, was the dynamo that would produce the electric current. A board containing a push button was set up in the execution room. This board held the switch that turns the current on and off, as well as a voltage meter to measure the exact amount of current used for the execution. A set of 24 lights were set up next to the electric chair to signal when the current was ready.

On August 6, 1890, Kemmler was fastened to Brown's electric chair, which was wired to an AC motor. When the current was switched on, it hit Kemmler so hard his fists clenched and his face contorted in agony. The current was turned off after about 17 seconds. Southwick, who was in attendance, quickly proclaimed that the electric chair was a success after ten years of work and research. However, Kemmler was not dead. He started gasping for air in front of the crowd of witnesses. It seemed that the technicians had not appropriately determined the amount of voltage required to kill a man.

The dynamo was switched back on, but it would take a while before it could build up its current. As electricity coursed through his body, Kemmler wheezed and shrieked in agony. His coat even caught fire.

It was minutes before Kemmler was finally dead. The scene was so gross that some of the horrified witnesses fainted while others retched. After confirming that Kemmler was dead, Dr. Edward Spitzka predicted that that would be the first and last death by electrocution. After hearing the reports of the execution, Westinghouse was horrified. He satirically stated that the state would have done a better job executing Kemmler with an axe.

Despite the failure of the electric chair execution at the Auburn prison, Edison and Brown did not give up. Edison told a reporter that he believed that in the future alternating current executions would run more smoothly. In a further attempt to convince the public that alternating current was more dangerous than direct current, Edison staged another demonstration in Coney Island, New York. The presentation was widely attended. In the demonstration, a circus elephant named Topsy would be executed. Topsy had been declared as too dangerous to people after killing three men. Edison had his team fit Topsy with copper wire sandals, and before thousands of individuals, Topsy was hit with 6000 volts of AC until she fell, dead.

Lighting The Chicago World Fair

The final battle in the war of currents was the race to light the Chicago World Fair. Also known as the World's Columbian Exposition, this was a Fair that was held to celebrate 400 years after the arrival of Christopher Columbus in the New World in 1492.

55

Chicago had emerged as the winner for the honors of hosting the World Fair, edging out other cities such as Washington D.C and New York City.

Before the fair, Chicago was considered a little backwater town in the Wild West with nothing much to offer. The Fair was the perfect opportunity for the city to gain some recognition on the national stage. Chicago was intent on proving itself once and for all by hosting an unforgettable Fair. Part of the strategy was to have the Fair lit by electricity, which was still considered a new phenomenon. Various electric companies had placed their bids to light the Fair.

Initially, Westinghouse and his company did not bid to supply electricity to the Fair. However, a small Chicago-based electric company had placed a low bid of $510,000. Since the company intended to use an AC-based system to power the Fair, they approached Westinghouse, and he agreed to act as a contractor for the local company. By this time, Edison's company had merged with the Thomson-Houston Electric Company to form a new company known as General Electric. General Electric had placed a $1,720,000 bid to supply the Fair with DC-based power. After going through the bids, the bid from General Electric was rejected for being too expensive. They submitted another $554,000 bid.

Despite all of Edison's previous efforts to demonstrate the dangers of alternating current, the contract to light up the Chicago World Fair was awarded to George Westinghouse. This was going to be a massive undertaking. Edison was very much angered by the fact that he lost the bid. In retaliation, he went to court over a patent dispute. Westinghouse intended to use Edison's one-piece incandescent bulb, which was more superior, to light up the Fair. Six months to the start of the World Fair, Edison won the case. Westinghouse could not use Edison's bulbs at the Fair.

Luckily, a couple of years early, Westinghouse had purchased the patents for another incandescent lamp from Sawyer-Man. He used this patent to side-step Edison's patent. Since Edison's design used a sealed globe of glass, Westinghouse came up with a design that used a ground glass stopper. Edison went to court to challenge Westinghouse's new design, but the courts upheld it as an independent design. With Edison out of the way, Westinghouse was now on course to light up the World Fair. With time running out, Westinghouse quickly manufactured 250 thousand of his newly designed bulbs.

On May 1, 1893, the Chicago World Fair opened its doors to the general public. On the evening of the opening day, President Grover Cleveland flicked a switch and about 100,000 incandescent lamps immediately illuminated the Fair's White City. Westinghouse's efforts had finally paid off immensely. This was the first ever large-scale test of the AC system. The visitors were awed by the lighting. The sheer beauty of a city illuminated by so many lights together was unlike anything the visitors had ever seen before. Each building was beautifully outlined using white bulbs. The Fair had turned into a demonstration piece for Westinghouse and his AC-based system.

The Chicago World Fair was a huge success. It ran for six months and attracted an attendance of over 27 million people. Westinghouse had finally showcased AC electricity not only to America but the entire world. He had shown how safe AC-based systems were. AC electricity was now going to spread from the East coast to the West coast and lands beyond. In addition to lighting up the city, Westinghouse had an exhibition showcasing many of Tesla's AC-based inventions, including motors, generators, and armatures. Losing the bid to light up the Chicago World Fair was the final nail in the coffin for Edison and his DC-based systems.

Harnessing The Niagara Falls

Lighting up the Chicago World Fair had given Westinghouse all the publicity he needed. He had also demonstrated the safety of AC electricity to the world. Shortly after the World Fair, the International Niagara Falls Commission called for proposals from experts on how to harness the roaring Niagara Falls to produce electricity. Set to be the largest hydro-electric power generation plant of its generation, this project was a product of sheer technological optimism. The project aimed to supply power to the industrial city of Buffalo, New York, which was 25 miles away from the generating plant.

After calling for proposals, the commission received several of them from various experts, ranging from systems using compressed air pressure to others using ropes, springs, and pulleys. Edison sent a proposal that would use DC to transmit power from the plant. However, the commission rejected all these proposals. At the time, the commission was headed by Lord Kelvin. Early on, Lord Kelvin had been very much opposed to AC electricity. After attending the Chicago World Fair, however, he got converted and became an active supporter of AC. Under his influence, the commission gave Westinghouse the contract to extract electricity from Niagara Falls using alternating current.

Ever since his childhood, Tesla's dream was to come to the United States and harness the great natural power of the mighty Niagara Falls. This dream became a reality when, in October 1893, the contract to build the power plant at Niagara Falls was awarded to George Westinghouse and Nikola Tesla. This was the last nail in the coffin for DC current. Under the able partnership of George Westinghouse and Nikola Tesla, the war of currents had been won. To add insult to Edison's defeat, Edison also lost his control over his company when J. P. Morgan – who had become the largest shareholder – instituted a merger between Edison General Electric and the Thompson-Houston Electric Company.

Initially, there were doubts that the plant at Niagara Falls would produce enough power to run the industries in Buffalo. No other power plant of such scale had ever been undertaken. However, Tesla was confident that Niagara Falls could produce enough electricity to power the whole eastern United States.

After a three-year construction period, which was filled with doubt and financial crises, the Niagara Falls hydroelectric power plant was completed in 1896. The switch at the power plant was flipped on November 16, transmitting the first 1000 horsepower of hydroelectricity to Buffalo. The demand for electricity among residents surged, the number of generators at the power plant multiplied, and power lines crisscrossed the city of New York.

With the completion of the Niagara Falls hydroelectric power plant, modern cities were born. Despite Edison's aggressive tactics to discredit AC electricity, its technical superiority was too much for the DC system to stand a chance. AC had won the war of currents to become the standard for the transmission of electricity. The use of DC electricity in the United States experienced a sharp decline. Edison would later admit that he was remorseful for not following Tesla's advice.

Chapter Six: The Aftermath of The Current Wars

After instituting the merger between Edison's company and Thompson-Houston, J. P. Morgan dropped the name Edison, resulting in a new company – General Electric. The new company shifted its focus from DC and directed its resources to AC. The new GE president and its board of directors silenced the opinions of Edison, who was still adamant about DC. Though Westinghouse had already gotten a head start in AC, General Electric was able to catch up with them in a few short years. The progress made by GE in AC is mainly credited to a Prussian mathematician by the name of Charles Proteus Steinmetz, who came to the company with a solid understanding of AC current from a mathematical perspective. The company hired several other engineers to improve the designs of its AC devices.

Despite having lost the war of currents, direct current electricity did not get relegated to museums of science and technology. Some cities continued to use the installed DC grids well into the 20th century. Some streets in Boston, Massachusetts continued to use 110 volts of direct current until the 1960s. Being the battleground for the war of currents, New York City had made massive investments in Edison's DC installations. These were not immediately decommissioned. They remained in operation for several years alongside the expanding AC network.

It was in 1927 that the state started systematically replacing the DC equipment. Despite this, the use of DC installations continued in some areas. The New York Hotel, which was built in 1929, had its own DC power plant. It continued to use the plant until the mid-1960s when it converted to alternating current. In 1998, Consolidated Edison started eliminating its remaining DC installations. By this time, the company had 4,600 DC customers. By 2006, 60 customers were still using the DC service. The last DC installation in the city was removed in November 2007 – over a century after the end of the Current Wars. However, some customers still have DC installations in their buildings. These are provided with on-site rectifiers that convert AC to DC.

The use of DC in other countries also survived well into the 20th century. In Helsinki, the DC network remained in operation up to the late 1940s, while in Stockholm, the last DC installation was decommissioned in the 1970s. In the United Kingdom, the Central Electricity Generating Board kept the Bankside Power Station on the River Thames – which produced 200 volts of DC electricity – in operation until 1981. The generating station provided power to the printing machinery in Fleet Street, which depended on DC electricity. The station was only decommissioned after the newspaper printing industry relocated to the docklands developing downstream, where it started using modern equipment powered by AC electricity.

Today, the situation is changing once again. Over a century after losing out to AC, DC electricity has started fighting back, this time, not through propaganda but its own merits. While AC was perfectly suited for the conditions of the 1880s and much of the 20th century, the power needs of the 21st century have started showing that it too has its limits. The most ironic bit is that direct current began mounting its comeback only a decade after New York City decommissioned its last DC installation.

The rise of High Voltage Direct Current (HVDC) has made it possible and even economical to transmit electricity using DC systems. HVDC systems were made possible by electronic devices like IGBTs, thyristors and mercury-arc valves, which were not available during the Currents War era. Today, several countries rely on HVDC systems for the bulk transmission of electric power from the distant power generating plants or for linking together separate AC systems. With HVDC systems, power still has to be converted from AC before being fed to the HVDC link. The main advantages of HDVC systems are that they enable higher power ratings for given lines and make it possible to transfer electrical energy between two unsynchronized AC systems. Already, countries like Brazil, Russia, India, and China have turned to HVDC systems as an alternative to AC systems for high-load transmission over long distances.

Today, the world is also experiencing a revolution in the manner in which electric power is produced and consumed. A huge percent of the world's electric power is generated from renewable energy sources (such as hydroelectric plants and offshore wind farms) that are located far from where the energy is needed. HVDC is now providing a far more efficient means of transmission compared to AC. Local power sources such as solar panels also produce DC electricity.

When it comes to power consumption, more and more equipment and appliances today depend on direct current. Our demand for consumer technology has had a significant impact on the recent comeback of DC systems. A direct current can only power any device or appliance that uses transistors. These include devices like personal computers, mobile communication gadgets, UPS systems, televisions and other electronic devices. These devices account for about 20 percent of the total power consumption in the world. Currently, these devices are fitted with their own rectifiers, which convert power from AC to DC. In the future, it may be possible to have large rectifiers that convert AC to DC before it enters each building, enabling these devices to work without the need of another rectifier. It's estimated that using DC rather than AC in buildings might result in 10 – 20 percent power savings.

Another thing that is driving the reemergence of DC electricity is the burgeoning dependence on data centers, which act as the backbone of today's digital world. These data centers host anywhere from tens to thousands of servers, all of which are run by DC electricity. Data centers are a very important part of most enterprises today – not just for tech companies, but for companies in a wide range of sectors, from research to finance to consulting. One thing that data centers are notorious for is that they are huge consumers of energy.

For instance, in Chicago, there is the Lakeside Technology Center, which is one of the world's largest data centers. This facility requires over 100 megawatts, making it the second largest consumer of energy in the region, coming in only after O'Hare International Airport. According to a report released by Lawrence Berkeley National Laboratory (LBNL), data centers within the United States consume over 14.6 terawatt-hours of electricity each year.

There are a couple of factors that make these data centers inviting targets for DC electricity. Since server downtime translates to huge losses for businesses, many data centers have uninterruptible power supplies (UPS) as power backups. This ensures that the servers continue running even in the event of a power outage. These UPS store power in the form of DC. This means that they have to convert it from AC to DC and then back to AC when supplying the data center. From there, this power has to be converted again to DC before it can be used by the servers since they cannot use AC.

This process is very inefficient and wasteful. As these conversions take place, some energy is lost in the form of heat. The heat is so much that the server rooms have to be equipped with specialized, energy-intensive cooling systems. These cooling systems sometimes require two times as much power as is needed by the servers themselves. All this wastefulness could be alleviated by powering data centers with DC electricity. Research conducted by LBNL found that DC data centers are 5% - 7% more efficient than AC data centers in terms of energy consumption.

Another factor that is helping DC gain traction is the proliferation of renewable sources of energy, such as wind farms and solar panels. These sources of energy produce DC electricity, which then has to be converted to AC before it can be used in homes and buildings, before being converted back to DC for use by electrical appliances. If you are trying to achieve a net-zero energy building – one whose energy production and consumption are equal – it makes sense to bypass the conversion to AC. As renewable energy sources become more common, DC systems will become cheaper, and we might start seeing buildings with both AC and DC systems, creating a new kind of hybrid buildings.

DC systems are also dominant in other applications that require relatively small amounts of electricity, such as:

- The electrical systems in vehicles – lighting, starting and ignition systems.
- Power installations that are not connected to the main grid, such as the power used in a ship.
- Utility-scale battery systems.
- Hybrid cars and other all-electric vehicles that rely on internal power supply for propulsion.
- Telecommunication plants.

Despite the recent resurgence of DC electricity, a worldwide transformation of the power system to DC will not take place anytime soon. Most countries have invested heavily in AC infrastructure. Changing these AC networks to DC would present a logistical nightmare. However, there is no denying the fact that more and more systems are being powered by DC today. DC has also started offering some key advantages that give it an edge over AC. The question that remains is, after losing the first War of Currents, is it possible for DC to fight back and maybe win another battle with AC? We can only wait and see.

Conclusion

The introduction of electricity as a source of lighting and power brought with it two competing transmission systems, which pitted Thomas Edison against George Westinghouse in a commercial rivalry, as each sought to make his system the standard. This rivalry saw Edison wage an aggressive and deceitful legal, media and PR campaign in an attempt to discredit Westinghouse's AC system. Despite Edison's propaganda, Westinghouse – with lots of assistance from Nikola Tesla – finally emerged the winner in the feud. At the end of the War of Currents, the AC system demonstrated its technical advantages and was adopted as the standard system of electricity transmission.